World Water Day 2023

Water is very important for us

Q. BANKS

Table of Contents

Introduction

World Water Day is a yearly occasion celebrated on Spring 22nd to bring issues to light about the significance of freshwater and promoter for the practical administration of freshwater assets. This day was first seen in 1993 after the Unified Countries General Get together proclaimed Walk 22nd as World Water Day.

Every year, World Water Day centers around a particular subject connected with water. The subject for 2023 is "Esteeming Water," which features the significance of perceiving the various upsides of water and what its shortage means for various areas of society.

The day is seen through different exercises and occasions like gatherings, courses, studios, and far-reaching developments. It gives an open door to people, associations, and state run administrations to meet up and make a move to address the difficulties of water the executives and guarantee the reasonable utilization of freshwater assets.

Notwithstanding the festivals on World Water Day, the Unified Countries has assigned the period from 2018 to 2028 as the Global 10 years for Activity on Water for Practical Turn of events. This decade intends to speed up progress towards the Practical Improvement Objectives connected with water and sterilization, and to address the difficulties connected with water the board and shortage.

Water Day in America

World Water Day is seen in America, alongside the remainder of the world, on Spring 22nd every year. The day is a chance for Americans to ponder the significance of freshwater assets and bring issues to light about the requirement for feasible administration rehearses.

In the US, World Water Day is praised through a scope of exercises, including instructive occasions, local area tidy up ventures, and strategy discussions. These occasions unite people, associations, and government organizations to examine the difficulties confronting water the executives and distinguish arrangements that can assist with guaranteeing a manageable future for freshwater assets.

Numerous American urban communities and states likewise utilize World Water Day as a valuable chance to send off water preservation and schooling efforts, empowering occupants to make a move to diminish their water utilization and safeguard nearby water sources. These missions frequently incorporate ways to preserve water at home, as well as data about water quality and the significance of safeguarding normal environments.

By and large, World Water Day gives a significant open door to Americans to meet up and zero in consideration on the basic issues connected with water the executives and maintainability. Through instruction, cooperation, and activity, Americans can pursue a future where freshwater assets are secured and monitored for a long time into the future.

water day in united states

In the US, World Water Day is commended on Spring 22nd every year. This day is a chance for Americans to consider the significance of freshwater assets and bring issues to light about the requirement for supportable water the board rehearses.

There are numerous ways that Americans notice World Water Day. A few associations hold instructive occasions or local area tidy up tasks to bring issues to light about water preservation and the significance of safeguarding our water assets. Different gatherings center around backing and strategy conversations, attempting to advance economical water the board rehearses at the nearby, state, and public levels.

Numerous urban areas and states likewise utilize World Water Day as an amazing chance to send off water protection crusades or give assets to occupants to get familiar with water preservation. These missions might incorporate ways to preserve water at home, data about water quality and water framework, and ways of engaging in neighborhood water preservation endeavors.

Generally, World Water Day in the US gives a stage to people, associations, and legislatures to meet up and zero in on the difficulties and open doors connected with water the executives and manageability. By cooperating to moderate and safeguard our water assets, Americans can assist with guaranteeing a supportable future for a long time into the future.

water day in Pakistan

In Pakistan, World Water Day is seen on Spring 22nd every year, alongside the remainder of the world. The day gives a potential chance to bring issues to light about the significance of freshwater assets and the requirement for supportable water the executives rehearses.

Pakistan is a country that faces huge water difficulties, including water shortage, unfortunate water quality, and wasteful water use rehearses. Thus, World Water Day in Pakistan is a significant occasion, uniting government offices, non-administrative associations, and people to talk about these difficulties and recognize arrangements.

On World Water Day in Pakistan, a scope of exercises happen, including courses, studios, and local area occasions. These occasions center around subjects like water protection, water quality, and water foundation advancement. They give an open door to partners to share information and thoughts and work towards further developing water the executives rehearses in the country.

Numerous associations additionally utilize World Water Day in Pakistan as a chance to send off water protection missions and bring issues to light about the significance of dependable water use. These missions might incorporate instructive materials, virtual entertainment missions, and local area outreach endeavors.

Generally speaking, World Water Day in Pakistan is a significant occasion that assists with bringing issues to light about the basic water difficulties confronting the country. By cooperating to advance maintainable water the board rehearses, Pakistan can guarantee that its freshwater assets are safeguarded and preserved for people in the future.

water day in uk

In the UK, World Water Day is seen on Spring 22nd every year, close by the remainder of the world. The day is a chance to bring issues to light about the significance of freshwater assets and the requirement for economical water the executives rehearses.

In the UK, World Water Day is commended through a scope of exercises, including instructive occasions, local area tasks, and strategy conversations. These occasions unite government offices, non-administrative associations, and people to talk about water-related issues and distinguish arrangements.

Numerous associations in the UK additionally utilize World Water Day as a chance to send off water preservation crusades or give assets to people and networks to become familiar with water protection. These missions might incorporate ways to monitor water at home, data about water quality and water framework, and ways of engaging in nearby water protection endeavors.

One of the central questions confronting the UK's water the board framework is the rising interest for water because of populace development and changing atmospheric conditions. World Water Day in the UK gives a chance to examine these difficulties and distinguish ways of advancing mindful water use and safeguard freshwater assets.

Generally, World Water Day in the UK is a significant occasion that assists with bringing issues to light about the basic water difficulties confronting the country. By cooperating to advance economical water the board rehearses, the UK can guarantee that its freshwater assets are safeguarded and moderated for people in the future.

water day in london

In London, World Water Day is seen on Spring 22nd every year, close by the remainder of the world. The day gives a chance to bring issues to light about the significance of freshwater assets and the requirement for reasonable water the board rehearses.

In London, a scope of exercises happen on World Water Day, including instructive occasions, local area ventures, and strategy conversations. These occasions unite government offices, non-administrative associations, and people to talk about water-related issues and distinguish arrangements.

Numerous associations in London additionally utilize World Water Day as a chance to send off water preservation crusades or give assets to people and networks to get familiar with water protection. These missions might incorporate ways to preserve water at home, data about water quality and water framework, and ways of engaging in nearby water preservation endeavors.

One of the major questions confronting London's water the executives framework is the maturing foundation, which can bring about holes and water wastage. World Water Day in London gives a valuable chance to examine these difficulties and recognize ways of advancing mindful water use and safeguard freshwater assets.

By and large, World Water Day in London is a significant occasion that assists with bringing issues to light about the basic water difficulties confronting the city. By cooperating to advance maintainable water the executives rehearses, London can guarantee that its freshwater assets are safeguarded and monitored for people in the future.

water day in dubai

In Dubai, World Water Day is seen on Spring 22nd every year, close by the remainder of the world. The day gives a chance to bring issues to light about the significance of freshwater assets and the requirement for supportable water the executives rehearses.

Dubai is situated in a district that faces critical water difficulties, including water shortage and unfortunate water quality. Accordingly, World Water Day in Dubai is a significant occasion, uniting government offices, non-administrative associations, and people to examine these difficulties and recognize arrangements.

On World Water Day in Dubai, a scope of exercises occur, including courses, studios, and local area occasions. These occasions center around points like water protection, water quality, and water framework improvement. They give an open door to partners to share information and thoughts and work towards further developing water the executives rehearses in the city.

Numerous associations in Dubai additionally utilize World Water Day as a chance to send off water preservation missions and bring issues to light about the significance of dependable water use. These missions might incorporate instructive materials, web-based entertainment missions, and local area outreach endeavors.

By and large, World Water Day in Dubai is a significant occasion that assists with bringing issues to light about the basic water difficulties confronting the city. By cooperating to advance supportable water the board rehearses, Dubai can guarantee that its freshwater assets are safeguarded and moderated for people in the future.

water day in china

In China, World Water Day is seen on Spring 22nd every year, alongside the remainder of the world. The day gives a chance to bring issues to light about the significance of freshwater assets and the requirement for reasonable water the executives rehearses.

China is a country that faces critical water difficulties, including water shortage, water contamination, and wasteful water use rehearses. Therefore, World Water Day in China is a significant occasion, uniting government offices, non-legislative associations, and people to examine these difficulties and recognize arrangements.

On World Water Day in China, a scope of exercises occur, including classes, studios, and local area occasions. These occasions center around themes like water preservation, water quality, and water framework advancement. They give an open door to partners to share information and thoughts and work towards further developing water the board rehearses in the country.

Numerous associations in China likewise utilize World Water Day as a chance to send off water preservation missions and bring issues to light about the significance of mindful water use. These missions might incorporate instructive materials, virtual entertainment missions, and local area outreach endeavors.

Generally, World Water Day in China is a significant occasion that assists with bringing issues to light about the basic water difficulties confronting the country. By cooperating to advance supportable water the executives rehearses, China can guarantee that its freshwater assets are safeguarded and moderated for people in the future.

water day in brazil

In China, World Water Day is seen on Spring 22nd every year, alongside the remainder of the world. The day gives a valuable chance to bring issues to light about the significance of freshwater assets and the requirement for reasonable water the board rehearses.

China is a country that faces huge water difficulties, including water shortage, water contamination, and wasteful water use rehearses. Therefore, World Water Day in China is a significant occasion, uniting government offices, non-administrative associations, and people to examine these difficulties and distinguish arrangements.

In Brazil, World Water Day is seen on Spring 22nd every year, close by the remainder of the world. The day gives a chance to bring issues to light about the significance of freshwater assets and the requirement for economical water the board rehearses.

Brazil is a country that faces huge water difficulties, including water shortage, water contamination, and lacking water framework in certain districts. Thus, World Water Day in Brazil is a significant occasion, uniting government offices, non-legislative associations, and people to examine these difficulties and distinguish arrangements.

On World Water Day in Brazil, a scope of exercises occur, including courses, studios, and local area occasions. These occasions center around points like water protection, water quality, and water framework improvement. They give an open door to partners to share information and thoughts and work towards further developing water the executives rehearses in the country.

Numerous associations in Brazil likewise utilize World Water Day as a chance to send off water protection missions and bring issues to light about the significance of dependable water use. These missions might incorporate instructive materials, online entertainment missions, and local area outreach endeavors.

In general, World Water Day in Brazil is a significant occasion that assists with bringing issues to light about the basic water difficulties confronting the country. By cooperating to advance economical water the executives rehearses, Brazil can guarantee that its freshwater assets are safeguarded and preserved for group of people yet to come

The end

www.ingramcontent.com/pod-product-compliance
Lightning Source LLC
Chambersburg PA
CBHW071147220526
45467CB00015B/2075